A cave is a big hole. It can be in a hill. We can go into this cave.

Caves are dim and damp.
You must be safe. You should
have a lamp.

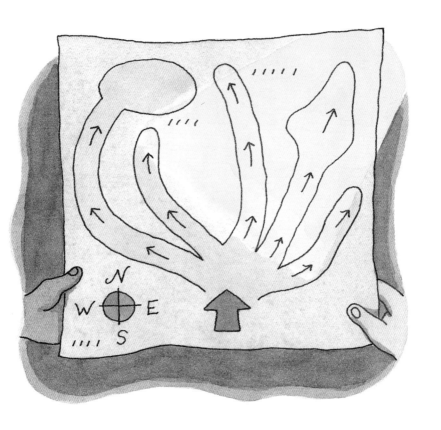

A cave has lots of paths.
Look at the map. This map
has five paths.

Pick one path. Take a rope
with you. Give the rope to your
friend.

A cave can have a lake.

Fish can live in the lake.

You can dive in the lake.

Bats and insects live in
caves. A bat does not see well.
But it can swoop and get bugs.

A cave can be wet. Some
rocks drip, drip, drip. Some
rocks look long, thin, and wet.

Your cave trip is done. Thank
the one who gave some facts.
Did you have a good time?